discovermore

A Look at Weather and Climate

Weather Instruments

Marie Harts

Britannica Educational Publishing

IN ASSOCIATION WITH

THE ROSEN PUBLISHING GROUP

Published in 2025 by Britannica Educational Publishing (a trademark of Encyclopædia Britannica, Inc.) in association with The Rosen Publishing Group, Inc.
2544 Clinton Street, Buffalo, NY 14224

Distributed exclusively by Rosen Publishing.
To see additional Britannica Educational Publishing titles, go to rosenpublishing.com.

Editor: Brianna Propis
Book Design: Michael Flynn

Photo Credits: Cover Zsolt Biczo/Shutterstock.com; (series background) Dai Yim/Shutterstock.com; p. 5 (top) PeopleImages.com - Yuri A/Shutterstock.com; p. 5 (bottom) Joel_420/Shutterstock.com; p. 6 Ilia Baksheev/Shutterstock.com; p. 7 Monkey Business Images/Shutterstock.com; p. 8 Image Point Fr/Shutterstock.com; p. 9 Alan.P/Shutterstock.com; p. 11 (top) Arnav Pratap Singh/Shutterstock.com; p. 11 (bottom) sumroeng chinnapan/Shutterstock.com; p. 12 Viacheslav Lopatin/Shutterstock.com; p. 13 Brester Irina/Shutterstock.com; p. 14 Alessiorez/Shutterstock.com; p. 15 Ryzhkov Oleksandr/Shutterstock.com; p. 17 (top) John D Sirlin/Shutterstock.com; p. 17 (bottom) Arthorn Saklang/Shutterstock.com; p. 18 Edward Haylan/Shutterstock.com; p. 19 Edward Haylan/Shutterstock.com; p. 20 Frank Wagner/Shutterstock.com; p. 21 https://commons.wikimedia.org/wiki/File:2008-07-09_Ely_Airport_ASOS_Ceilometer_in_Ely,_Nevada.jpg; p. 22 Christopher Wood/Shutterstock.com; p. 23 Farofang/Shutterstock.com; p. 24 Andrei Armiagov/Shutterstock.com; p. 25 flashpict/Shutterstock.com; p. 26 Burben/Shutterstock.com; p. 27 elRoce/Shutterstock.com; p. 29 (main) J.J. Gouin/Shutterstock.com; p. 29 (inset) Akimov Igor/Shutterstock.com.

Cataloging-in-Publication Data

Names: Harts, Marie.
Title: Weather instruments / Marie Harts.
Description: New York : Britannica Educational Publishing, in Association with Rosen Educational Services, 2025. | Series: Discover more! a look at weather and climate | Includes glossary and index.
Identifiers: ISBN 9781641901383 (library bound) | ISBN 9781641901376 (pbk) | ISBN 9781641901390 (ebook)
Subjects: LCSH: Meteorological instruments--Juvenile literature. | Weather--Juvenile literature.
Classification: LCC QC876.W347 2025 | DDC 551.5028'4--dc23

Manufactured in the United States of America

CPSIA Compliance Information: Batch #CSBRIT25. For further information contact Rosen Publishing at 1-800-237-9932.

Contents

What's the Weather Today?

It is very likely you've witnessed a conversation about weather, or perhaps you've had one yourself. Most people make casual remarks about the weather, and that is enough for most of us. However, meteorologists (scientists who study the weather) want to know precisely how hot it is, how much it is raining, and from what direction the wind is blowing. Is it really freezing or is it just very cold? Meteorologists use many instruments to answer these questions. Instruments are important for meteorologists to understand and predict the weather, and you'll be able to describe the weather like them after learning about their instruments!

compare and contrast

Meteorologists study the daily changes in weather in a particular area. Climatologists study the average weather conditions over long periods of time all over Earth. How might their jobs be similar? How might they be different?

Watching or feeling raindrops fall lets you know that it's raining outside, but meteorologists need a special instrument to tell them how much rain is falling.

Have you ever noticed this on top of somebody's house? It tells them which direction the wind is blowing!

Thermometers

The thermometer is one of the oldest weather instruments created—and we still use it today! It measures temperature, or how hot or cold it is. The first thermometer was invented by an Italian scientist named Galileo in about 1592. Many different types of thermometers have been developed since then. The most basic type consists of a glass tube with liquid, called **mercury**, in it. Mercury expands when it heats up and shrinks when it cools down. These changes make the mercury level in the tube rise and fall. Markings on the outside of the tube show the degrees on a temperature scale.

Not only did Galileo invent the thermometer, but he also used a telescope to make discoveries about space.

A thermometer can also be used when people are feeling sick. It tells you the temperature of your body.

There are several different scales that measure temperature. The two most common are Fahrenheit and Celsius. The Fahrenheit scale is used in the United States for everyday measurements. On this scale, the freezing point of water is 32 degrees and the boiling point is 212 degrees. The Celsius scale is used for scientific work and everyday weather measurements in much of the rest of the world. The freezing point of water is 0 degrees and the boiling point is 100 degrees on this scale.

WORD WISE

MERCURY IS A SILVERY-WHITE METAL THAT IS LIQUID AT ORDINARY TEMPERATURES.

Weight of the Air

There is a layer of gas that surrounds and protects Earth, known as the atmosphere. It is often called air. The weight of the air is called atmospheric, or barometric, pressure. Meteorologists use barometers to measure this pressure. The atmosphere is always moving and changing, and atmospheric pressure changes with it. These changes are associated with the weather. Following the changes in atmospheric pressure helps meteorologists predict changes in weather. If the pressure is falling, stormy weather is probably on the way. Rising pressure usually means the return of nice weather. If the pressure remains steady, it means the weather will likely stay the same.

Barometers can alert meteorologists of thunderstorms. They are caused by changes in atmospheric pressure.

The barometer was invented by Evangelista Torricelli. He was a student of Galileo's before Galileo died.

Consider This

What other ways might meteorologists use to predict incoming storms?

Humidity

Humidity is the amount of water vapor and moisture in the air. It is the element of the atmosphere that changes most frequently. Meteorologists measure humidity with a device called a hygrometer. Simple hygrometers measure humidity by using a strand of human hair attached to a spring and dial. As the humidity increases, the hair gets shorter. This action pulls the spring and makes the dial hand move. When the humidity is lower, the hair is longer. This reaction loosens the tension on the spring and allows the dial hand to move in the other direction. Meteorologists use humidity levels to predict when it will rain or snow.

Consider This

Relative humidity is the amount of moisture in the air compared to the highest amount of moisture possible at the same temperature. A hot desert will have almost no relative humidity. However, a rainforest at the same temperature can have a very high relative humidity. Why do you think a rainforest would have a high relative humidity, whereas a desert at the same temperature would not?

Some places around the world have very high humidity. The most humid city in the world is Patna, India, with a humidity average of 99.2 percent.

The three main factors impacting changes in humidity are air temperature, warm bodies of water, and air movement.

Here Comes the Rain Again

The amount of rain that falls in a given time or place is measured with a rain gauge. Rain gauges are open-mouthed containers that catch rain as it falls. A typical rain gauge is made up of a measuring tube with a funnel on top. Meteorologists can tell how much it has rained by comparing the water level with the measurement marks on the container.

The rain gauge is one of the most ancient weather instruments. The ancient Greeks and people in India kept rainfall records around 500–400 BCE.

Mount Waialeale, on the island of Kauai, Hawaii, has an average yearly rainfall of 384 inches (9,763 millimeters). It is known as one of the wettest places on Earth!

Another type of rain gauge is the tipping bucket rain gauge. This gauge automatically empties itself as the rain is measured. Knowing how much rain falls in an area helps farmers decide what crops to grow and when to plant and harvest them. If an area is getting too much or too little rain, scientists can measure this with rain gauges.

Consider This

Why would it be important for farmers to understand their area's rainfall patterns?

Weather Vanes

Wind is an important element of weather, so studying it is crucial for meteorologists. The wind blowing around the world greatly affects climate and weather. Weather vanes (also known as wind vanes) tell us in which direction the wind is blowing. A simple weather vane is made of a piece of metal shaped like an arrow. The arrow turns until it faces the direction from which the wind is blowing. Winds are named for the direction from which they come. A wind that comes from the west is a westerly wind. A storm's movements can also be tracked using a weather vane.

Weather vanes have been used for over 2,000 years. They were invented by the Greek astronomer Andronicus in 48 BCE. The first weather vane design was inspired by the Greek god Triton.

If you were to throw a piece of grass in the air on a windy day, the grass would act as a weather vane and fall toward the direction of the wind.

Consider This

Other than meteorologists, what jobs do people have where it's important for them to know the direction that the wind blows?

Wind Speeds

Wind speeds are measured by anemometers. They look and work similar to weather vanes. Many anemometers have wind vanes attached to them. An anemometer is made of three or four cups attached to the ends of arms. When the wind blows, it makes the anemometer spin. The faster the wind speed, the faster the anemometer turns. Meteorologists use anemometers to measure the strength of storms. Tropical storms have wind speeds of 39 miles (63 km) per hour or higher. A tropical storm becomes a hurricane when the wind speed goes over 74 miles (119 km) per hour. Tornadoes reach wind speeds as high as 300 miles (483 km) per hour.

Consider This

Weather vanes and anemometers both measure the wind, but they focus on different parts. Why is it important to know both wind direction and speed?

The violent winds of a tornado rotate and are funnel-shaped. They extend from a thunderstorm to the ground and can lift objects, such as cars or animals, into the spinning air.

The cups on an anemometer catch the wind, and the arms are attached to a vertical rod. As the wind blows, the cups rotate and cause the rod to spin. The anemometer counts the number of rotations, or turns, which is used to calculate wind speed.

Balloons

Balloons aren't just found at birthday parties! Weather balloons were originally used to study wind patterns. A balloon was sent into the air and carried by wind currents. People on the ground used telescopes to study the speed and direction of the balloon at different **altitudes**. Meteorologists eventually wanted more information on conditions higher up in the atmosphere. This desire led to the invention of the radiosonde in the late 1930s.

The higher a weather balloon travels, the more it expands, typically from 6.5 feet (2 m) up to 26 feet (8 m).

A radiosonde is a collection of instruments that is attached to a weather balloon. As the balloon rises, the radiosonde records changes in air pressure, temperature, and humidity. This information is sent to the ground by radio. The balloon bursts when it reaches about 100,000 feet (30,480 meters) high. Then, the radiosonde is carried to the ground by a parachute.

WORD WISE

ALTITUDE IS THE HEIGHT OF SOMETHING ABOVE SEA LEVEL.

Clouds

Millions of tiny water droplets or ice crystals combined with dust particles floating together in the air are known as clouds. Meteorologists use ceilometers to measure the height of clouds. Ceilometers also measure the thickness of clouds. They shine a beam of light up at the clouds. The light is reflected from the clouds back to the ceilometer. Ceilometers use special telescopes to detect this reflection. Ceilometers can take measurements in the daytime or at night. Clouds often signal a weather change. Rising cloud levels mean the weather is clearing. Rain is coming when there are lower and thicker clouds in the sky.

The three main types of clouds are cirrus (high and thin), cumulus (puffy), and stratus (layered). Which one is seen here?

A ceilometer shoots a laser beam up into the sky at the base of clouds.

Consider This

Airports use ceilometers to measure the base height of clouds. Why would an airport need to know the height of clouds?

Advances in Technology

Developed in the early 1800s, the telegraph was the fastest way to communicate over long distances during that time. It used electricity to send coded messages through wires. It allowed scientists to collect and analyze data from many regions on Earth, including the oceans, deserts, and rainforests.

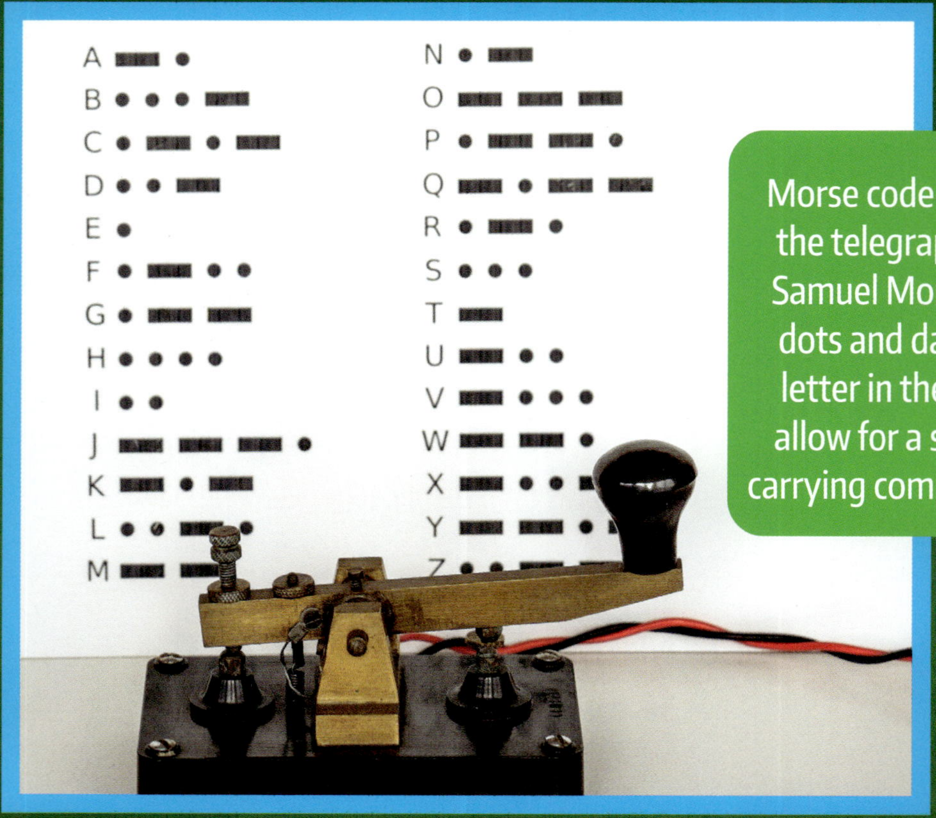

Morse code—named after the telegraph's inventor, Samuel Morse—assigned dots and dashes to each letter in the alphabet to allow for a simple way of carrying complex messages.

Today, the Internet and cell phone apps allow for weather communication to be as fast as ever. Anyone can access forecasts for current and upcoming weather from their personal devices.

In the 1900s, computers became valuable tools in meteorology. They replaced the telegraph for communicating information about weather. Computer programs also use mathematics to create models of weather patterns. These models can reliably forecast the weather—even one month ahead of time.

Consider This

The telegraph and computers made communication about the weather fast and efficient. Why is it important for people to know about the weather as soon as possible?

Satellites

In 1960, the National Aeronautics and Space Administration (NASA) launched the first weather **satellite** into space. Weather satellites give meteorologists information about the atmosphere. Meteorologists use weather satellites to photograph weather systems, such as storms, around the world. This ability makes it possible for meteorologists to see changes in weather as soon as the changes happen.

Satellites orbit Earth in outer space and send information back to meteorologists.

This is how a hurricane looks over the Atlantic Ocean from a satellite image.

Weather satellites provide photographs that make it possible to find and follow weather systems from the moment they form. Satellites also send lots of information on atmospheric conditions to scientists on the ground. Today, large storms are unlikely to strike without warning because weather satellites allow meteorologists to see them coming. This makes communication about the weather and people's understanding of it fast and advanced.

WORD WISE

SATELLITES CAN BE NATURAL OBJECTS OR MACHINES, LIKE WEATHER SATELLITES, MADE BY HUMANS. BOTH TYPES TRAVEL AROUND OUR PLANET OR OTHER OBJECTS IN SPACE.

Waves of Energy

Radar is a technology system that relies on waves of energy to sense objects. Radar can find a faraway object and tell how fast it is moving. Radar units send radio waves, which are invisible streams of energy, into the air. When the waves hit an object, they bounce back to the radar unit. By measuring how long it takes the waves to return, the radar can tell how far away the object is. Radar can also tell where a moving object is headed and at what speed.

Air traffic controllers also use radar to detect and display aircraft in the sky.

Radar can track tiny amounts of snow as well as intense weather like lightning storms.

Sometimes radar can tell how big objects are. Hurricanes, thunderstorms, tornadoes, and other powerful storms can each be detected by radar, which helps people to prepare for them.

Consider This

Aside from meteorologists and air traffic controllers, who else might use radar to track objects? What would they track them for?

You Can Do It Too!

Meteorologists use a variety of instruments to record, measure, and analyze the weather. But they are not the only people who use weather instruments. You can use an outdoor thermometer and barometer to keep track of the temperature and atmospheric pressure in your neighborhood. Home-improvement stores sell inexpensive hygrometers that you can use to measure the humidity right outside your window. And you can easily make a rain gauge, a wind vane, and an anemometer yourself! You may have some trouble getting your hands on a radiosonde or a ceilometer, but you can definitely attach a helium balloon to a long string and use it as a miniature weather balloon. You can even use the internet to see satellite images from space and track weather patterns on radar! But be wary of dangerous weather conditions, especially if there is lightning around.

Consider This

Why is it important for people to know about upcoming weather and talk about it with others?

Weather instruments can be used for science projects in school.

This rain gauge is being used to monitor the amount of water an onion garden is receiving.

Glossary

astronomer A person who studies space.
atmospheric pressure The weight of the air.
climate The long-term weather conditions of any place.
complex Having many different and complicated parts.
efficient Describes doing something well without wasting time.
funnel-shaped Wide at the top and narrow at the bottom.
hurricane An extremely large, powerful, and destructive storm with very strong winds.
instrument A device that measures something (such as temperature or distance).
meteorology The study of daily changes in weather in a particular place.
predict To make a guess about something that might happen.
tornado A storm in which powerful rotating winds form a column that reaches from a cloud down toward the ground.
Triton The Greek mythological god of the sea. He had the torso of a man and the tail of a fish.
tropical storm A powerful storm that begins over warm tropical oceans and has winds that are not as strong as those of a hurricane.
vertical Standing up and down.

For More Information

Books

Raij, Emily. *Meteorologists.* North Mankato, MN: Pebble, 2020.

Schuh, Mari C. *Measuring the Weather.* Vero Beach, FL: Rourke Educational Media, 2020.

Schwartz, Heather E. *Tracking a Storm.* Huntington Beach, CA: Teacher Created Materials, 2018.

Websites

National Geographic Kids: Make a Weather Station
kids.nationalgeographic.com/science/article/make-a-weather-station
Follow the steps to build your own weather station, filled with tools that meteorologists use.

National Geographic: Meteorology
education.nationalgeographic.org/resource/meteorology/
Learn more about the study and practice of meteorologists.

National Geographic: Storms
www.nationalgeographic.co.uk/storms
Explore helpful articles and recent headlines about storms.

Index